T·H·E B·E·S·T O·F
APPLES

JG
PRESS

4445
Published in the USA 1995 by JG Press
Distributed by World Publications, Inc.
Copyright © 1995 by CLB Publishing
Godalming, Surrey, UK

Printed and bound in Italy
ISBN 1-57215-043-2

The JG Press imprint is a trademark of JG Press, Inc.
455 Somerset Avenue
North Dighton, MA 02764

T·H·E B·E·S·T O·F
APPLES

CONTENTS

INTRODUCTION

Of all the fruits that grow in the temperate world, the apple is the undoubted king. The moist, cool climate seems to bring out all of its sharp flavor in a crisp, juicy texture, and the farther south one goes, the softer and mealier the apples become and the blander their sweetness.

The apple is thought to have originated in the mountains of the Caucasus and its cultivation goes back to the Stone Age, even farther if the story of Adam, Eve and the Tree of Knowledge is to be taken literally. Some think that it was not an apple tree at all that was involved but another fruit, possibly an apricot or a fig, but the majority still come down in favor of the apple.

The Ancient Egyptians cultivated apple trees and the Greeks and Romans knew them well. By the end of the fourth century AD, there were thirty-seven varieties of apple on record. From then on, wherever western civilization went, apples went too – to the New World, to the Antipodes as well as throughout the Old World, to settle wherever the climate and growing conditions suited it. Today, there are well over a thousand varieties of apple in existence but the search for perfection continues unabated. Commercial growers have still to find the perfect apple, one that is a heavy cropper and resistant to disease, of equally good flavor and appearance, unaffected by packing and lengthy transportation.

The apple's versatility cannot be overemphasized. It can be enjoyed at any time of day, in any course of a meal. A fine French cuisine, that of Normandy, relies heavily on apples – fresh, cooked with all kinds of meat, poultry and game, or sliced into large open tarts. The French also have a magnificent apple brandy called Calvados. Americans have Mom's apple pie, non-alcoholic apple cider, and a brandy called apple jack. Last but not least, there is cider vinegar, which is infinitely preferable to some harsh, inferior potions that go by the name of wine vinegar.

As far as the cook is concerned, apples fall into two main categories, sweet and sour. However, there is no reason why you should not explore the flavors and textures of sweet apples for savory sauces, pies and other apple dishes as well.

When choosing apples in the store or market, go for fresh, firm fruit with a smooth, shiny skin and good color. A pale, anaemic color is usually a sign of immaturity and consequent lack of flavor, which is unlikely to improve, no matter how long you wait. A slightly shrivelled skin could be caused by incorrect or overlong storage. An overripe apple, which also tends to be over-size, will be mealy and rather soft to finger pressure, and its flavor will be impaired.

Apples are best frozen as pie slices or purées or apple sauces. Blanch slices in acidulated water and dry or sugar freeze. The can also be dried.

Apples are not just for eating by themselves, for cooking with, or for making wine. They are also very good for you and 'an apple a day keeps the doctor away' is an old saw which is based on scientific fact. Apples, particularly the peel, are rich in vitamins and contain at least six essential minerals, as well as malic acid, which is a powerful aid to digestion. A raw apple is good for the teeth and gums and sweetens the breath; eaten last thing at night it will help you to sleep well. Apple pips can be poisonous, but only in considerable quantities – a cupful has been known to be lethal.

POLISH HERRING
APPETIZER

1 lb salted herrings
2 tablespoons, finely chopped mild onion or shallots
8 tablespoons coarsely chopped, peeled apple
8 tablespoons thick sour cream
2 tablespoons lemon juice
½ tablespoon sugar
1–2 tablespoons finely chopped parsely, chives or scallions to garnish

Desalt the herrings for 48 hours as follows. Place them in a large bowl in the kitchen sink and let cold water trickle over them constantly until required; or, alternatively, cover the herrings with cold water and leave for 48 hours, changing water frequently – several times a day.

Drain the herrings thoroughly. Fillet them, discarding all skin and bones, and cut each fillet into bite-sized pieces. Arrange the pieces in a shallow serving dish.

In a small bowl, mix the onion or shallots and apple with the sour cream, and flavor to taste with lemon juice and sugar. Pour over the pieces of herring and mix lightly but thoroughly to coat them completely. Scatter the top with parsley, chives or scallions and put aside in a cool place for a couple of hours to develop flavors before serving.

Serves 6

HOT APPLE SOUP

1½ lbs apples
One 2 inch piece of cinnamon stick
Twist of thinly peeled orange rind
1½ tablespoons potato flour or cornstarch
1¼ cups orange juice
½–¾ cup sugar
Lemon juice
1¼ cup hot cream
Ground cinnamon or cinnamon croûtons, to serve

Wash, quarter and core 1 lb of the apples. Put them in a pan with the cinnamon stick, orange rind and 2¾ cups water. Bring to the boil and simmer, uncovered, until the apples have disintegrated. Rub the mixture through a strainer, discarding the cinnamon stick and rind. Pour the soup back into the rinsed-out pan.

Blend the potato flour or cornstarch smoothly with a few tablespoonfuls of orange juice. Reheat the soup and when it is halfway to the boil, stir in the flour paste, together with the remaining orange juice. Sweeten to taste with sugar, bring to the boil, stirring, and simmer gently for 3 or 4 minutes.

Peel and core the remaining apples. Grate them into long shreds on the coarsest side of a grater. If the soup is not to be served immediately, squeeze a little lemon juice over them and toss lightly with a fork.

Just before serving, bring the soup to simmering point. Stir in the grated apples and cream. Taste for sweetness and serve, each portion garnished with either a pinch of cinnamon or a few croûtons.

Serves 4

BAKED CHICKEN WITH APPLES

1 roasting chicken, about 3½ lbs
¼ cup butter
3 slices lean bacon, diced
1½ lbs firm, aromatic apples
Salt and freshly ground black pepper
1¼ cups dry apple wine or cider

Clean the chicken carefully and cut it into 8 or 9 neat pieces. Dry the joints with paper towels and put them aside until needed.

In a large, heavy frying pan, melt the butter until foaming but not colored. Add the diced bacon and fry gently until limp and golden but not crisp. With a slotted spoon, drain bacon bits and put aside on a plate until needed.

In the fat remaining in the pan, sauté the chicken joints over moderate heat until a rich golden color on all sides. Quarter, core and slice the apples vertically. (I prefer not to peel them as the skin helps the slices to keep their shape).

In a heavy, ovenproof casserole with a tight-fitting lid, arrange a third of the apple slices. Lay half of the chicken pieces on top and scatter with half of the bacon pieces. Season lightly with salt and freshly ground black pepper. Repeat with another layer of apples, the remaining chicken joints and bacon pieces, season lightly and top neatly with the remaining apple slices.

Pour the cider into the juices remaining in the frying pan, and, over a moderate heat, scrape the bottom and sides of the pan clean with a wooden spoon. Pour the simmering liquid all over the top layer of apples. Cover the casserole tightly and bake in a moderate oven 375°F, for 40–45 minutes, or until the chicken is tender. Serve straight from the casserole.

Serves 6

PERSIAN STUFFED APPLES

8–10 large, tart apples
Butter
Ground cinnamon (optional)

Stuffing
⅔ cup yellow split peas
1 large onion, finely chopped
2 tablespoons butter or oil
1 lb lean ground beef
½ teaspoon ground cinnamon
Salt and freshly ground black pepper

Basting sauce
4–6 tablespoons cider vinegar
2–3 tablespoons sugar

Start by preparing the stuffing. Boil the split peas in unsalted water until they are very soft, about 30 minutes. Meanwhile sauté the onion in butter or oil until soft and lightly colored. With a slotted spoon, transfer onion to a plate and in the same fat sauté beef until it changes color. Return onion to the pan and mix well.

When split peas are soft and almost mushy, drain them thoroughly and mix them with the beef and onion. Sprinkle in the cinnamon and season to taste with salt and freshly ground black pepper. Leave to cool.

Meanwhile, prepare the apples. Wipe them clean with a damp cloth and core them from the stem end to within about ½ inch of the bottom. Core out some of the pulp as well, leaving thick shells firm and unbroken. Chop up the pulp.

Lightly butter a baking dish that will take all the apples in a single layer. Scatter the apple pulp evenly over the dish. Sprinkle with a pinch of cinnamon, if liked, and moisten with a few tablespoons of water.

Stuff the apples with the meat mixture and arrange them side by side on the chopped apple pulp. Dot each apple with a flake of butter. Bake in a moderate oven, 350°F, for 30 minutes.

Prepare a sweet-sour basting sauce for the apples by combining the vinegar and sugar with ½ cup water. Bring to boiling point, stirring until sugar has melted, and simmer for 1 minute.

Spoon a little boiling sauce over each apple. (Any sauce left over may be poured over the chopped apples lining the baking dish.) Continue to bake apples for about 15 minutes longer until they are quite soft. Crush the apple pulp and pan juices together, and serve as a sauce for the apples. Serve hot.

Serves 4–5

WALDORF SALAD

4 red-skinned firm apples
3–4 tablespoons lemon juice
4–6 stalks celery
½ cup walnuts
½ cup raisins
⅔ cup thick mayonnaise

Quarter, core and cut the apples into small dice without peeling them. Toss them with lemon juice to prevent discoloration. Halve celery stalks lengthwise if they are thick and slice them thinly. Coarsely chop the walnuts and rinse the raisins, draining them thoroughly.
In a large mixing bowl, combine the apples and their juices with the remaining ingredients. Add mayonnaise and toss lightly until well mixed. Pile into a bowl and serve.

Serves 4–6

BULGUR PILAF WITH APPLES AND ALMONDS

1 large onion, finely chopped
¼ cup butter
3 cups bulgur wheat
3 tart, crisp eating apples, peeled, cored and diced
¼ teaspoon crumbled dried thyme
¼ teaspoon celery seed
1 tablespoon finely chopped parsely
2 cups well-flavored chicken stock
Salt and freshlyt ground black pepper

Almond garnish
8 tablespoons slivered blanched almonds
¼ cup butter

In a large, heavy saucepan or flameproof casserole with a tight-fitting lid, sauté onion in butter until soft and golden. Stir in dry bulgur and sauté for a few minutes longer over moderate heat until lightly colored and aromatic. Add diced apples and continue frying, stirring, until each cube is coated with butter. Sprinkle with dried thyme, celery seed and parsely, and stir in chicken stock.

Bring to boiling point, cover pan tightly and either cook over the lowest possible heat for about 25 minutes until all the stock is absorbed or bake in a moderate oven 350°F, for roughly the same length of time. When bulgur is cooked, fluff it up with a fork and correct seasoning with a little salt or pepper if necessary.

Have almonds already sautéed until golden in butter. Serve bulgur in a heated bowl with hot almonds and butter poured over the top.

Serves 4–6

SALADE PARISIENNE

2–3 large, crisp apples
2 large potatoes, boiled in their jackets
3 heads of chicory
2 bananas
1 large stalk celery
3 carrots
¼ cup nuts, coarsely chopped
¼ cup raisins

Dressing
4 tablespoons salad oil
1½ tablespoon lemon juice
Pinch of paprika
Salt and freshly ground black pepper

Make the dressing for the salad first. Put the oil, lemon juice, paprika, salt and black pepper to taste in a screw-top jar. Close tightly and shake well to emulsify dressing.

Peel, core and dice the apples, and put them in a large bowl. Peel and dice the potatoes, and add them to the apples. Trim the chicory heads. Slice them thinly, together with the bananas and celery, and add them to the bowl. Shred the carrots coarsely and add them. Finally, add the nuts and raisins.

Toss all the ingredients gently with a large kitchen fork until well mixed. Shake the dressing up once more and mix it lightly but thoroughly into the salad. Correct seasoning and chill lightly until ready to serve.

Serves 6

Note: If you find the salad is rather 'dry', make up some more dressing in the same proportions and add it to the bowl (the exact amount needed will depend on the juiciness of the fruit and vegetables).

ITALIAN APPLE PANCAKE

2½ tablespoons flour
Pinch of salt
⅔ cup milk
1 tablespoon sugar
½ teaspoon vanilla essence
3 tablespoons golden raisins
2 eggs
2 small apples
Butter, for frying
Sifted confectioners' sugar, to serve

First prepare a thin batter by blending the flour and salt smoothly with the milk. Stir in sugar and vanilla essence.

Rinse raisins and put aside to drain in a strainer. Beat eggs lightly in a large bowl and gradually beat in prepared batter. When smoothly blended stir in drained raisins. Put aside while you quarter, peel, core and slice apples paper-thin. Stir apples into the prepared batter.

Bring out a thick frying pan 7–8 inches in diameter and a flat plate that is slightly larger. Grease plate lightly all over with butter. In the frying pan, melt a large knob butter and when sizzling, tilt pan around so that entire surface is coated.

Pour in batter, making sure apple slices and raisins are evenly distributed in the pan. Cook over low heat until pancake batter has set on top and pancake is a rich golden brown underneath.

Carefully turn pancake out upside down on to the buttered plate. Melt a small knob of fresh butter in the pan and slide the pancake back in to cook and brown the other side over low heat.

Invert pancake on to a heated serving dish.

Dredge with confectioners' sugar and serve immediately, cut in wedges.

Serves 2 as a snack, 4 as a dessert course

SOUTHERN APPLE PIE

Pie dough made with 2 cups flour, 1 teaspoon salt, ⅔ cup
plus 2 tablespoons butter, for a 2-crust
9 in pie
1½–2 lbs tart apples
Lemon juice (optional)
½–¾ cup white or brown sugar, or half and half
¼ teaspoon ground cinnamon
⅛ teaspoon grated nutmeg
3 tablespoons plain flour
Pinch of salt
¼ cup seedless raisins
2 tablespoons butter
Sugar

Start by lining a pie dish with the rolled-out pastry,
overhanging the sides a little, and cutting out a lid to
go on top. Put aside until needed.
Quarter, peel and core apples, and slice them thinly
into a bowl. If they are not too sour, or are rather old
and tasteless, toss them with a tablespoon of lemon
juice.

Blend sugar with spices, flour and salt, and sprinkle over apples. Add raisins and toss lightly until thoroughly mixed.

Pack apple mixture into pastry-lined pie dish. Dot surface all over with flakes of butter.

Lay pastry lid in position. Press all around sides to seal them, then flute rim attractively between forefinger and thumb. Cut slits in top crust.

Bake pie in a hot oven 450°F, for 15 minutes, then reduce heat to moderate 350°F and continue to bake for 35–40 minutes longer, or until top crust is a rich golden color and apples feel soft when prodded with a skewer through one of the slits.

Serves 6–8

SHERRIED APPLES WITH SPICED NUTS

6 large baking apples
Butter
⅔ cups sherry
Whipped cream, to serve

Nut stuffing
¼ cup butter
Scant ⅔ cup brown sugar
1 teaspoon ground cinnamon
1 tablespoon sherry
8 tablespoons chopped walnuts

Prepare apples in usual way, coring them and peeling away about a third of the skin, starting from the top.
Prepare nut stuffing by creaming butter and brown sugar together until light. Beat in cinnamon and sherry until thoroughly blended, then stir in chopped nuts.
Fill apples with nut stuffing. Arrange them side by side in a buttered baking dish. Pour sherry evenly over the top.
Bake apples in a moderate oven, 375°F, basting frequently with pan juices, for 30–40 minutes, or until they feel soft when pierced lightly with a skewer.
Serve lukewarm or cold, with a bowl of whipped cream.

Serves 6

HOT APPLE CHARLOTTE

Butter and breadcrumbs for baking dish
12 slices bread
1¾ cups milk
2 eggs
1½ tablespoons sugar
8 apples, peeled, cored and sliced
3 tablespoons chopped candied peel
3 tablespoons chopped raisins
¼ cup confectioners' sugar, sifted
1 teaspoon vanilla essence

Butter a deep baking dish (a soufflé dish is ideal) and coat it with breadcrumbs. Toast the bread slices lightly on both sides until golden but not browned.

Dip each slice of toast on both sides in a mixture of milk beaten with eggs and sugar, and put them aside on a plate. Mix the apples with the candied peel, raisins, confectioners' sugar and vanilla.

Arrange four layers of toast and three layers of apples in the baking dish (ie, starting and ending with a layer of toast). Pour in any remaining milk mixture.

Bake the charlotte in a moderately hot oven, 400°F, for 45 minutes, or until the top is crisp and golden. Serve hot.

Serves 4–6

BAKED RICE AND APPLE PUDDING

1 cup short-grain rice
¼ cup butter
Salt
¼ cup seedless raisins
1–2 tablespoons chopped crystallized orange peel
2 lb crisp, tart apples
White or brown sugar
1 teaspoon vanilla essence

Measure the rice and put it in a pan. Add 3 times the volume (ie about 3 cups) of boiling water, half the butter and a generous pinch of salt; stir well and cook very gently with the lid half on the pan until the water has been absorbed and the rice is very soft. Stir in the raisins and peel, cover slightly and put aside while you prepare apples.

Peel, core and slice the apples very thinly into a bowl. Add sugar to taste, sprinkle with vanilla and toss well with a large fork.

Grease a large baking dish generously with some of the butter, and spread base with a third of the rice mixture. Cover with half of the apples, followed by half of the remaining rice, all the remaining apples (and their juices, if any) and the rest of the rice, spreading it out as evenly as possible with the back of your spoon.

Dot the top of the pudding all over with flakes of remaining butter and bake in a moderate oven, 350°F, for 45–50 minutes, or until surface is crusty and golden brown. Serve hot.

Serves 6

NEW YORK APPLE CANDY

---

12 apples
2 cups sugar
¼ teaspoon salt
2 tablespoons unflavored gelatin
1 cup chopped walnuts
1½ tablespoons lemon juice
Confectioners' sugar

Quarter apples, including the skins and cores. Put them in a heavy saucepan with 1 cup water, and cook over a low heat until very soft. Press through a fine sieve or colander, discarding the skins and cores as well as any stray pips.

Put the apple pulp, sugar and salt back in the cleaned pan and cook over a low heat until very thick, stirring constantly; about 15 minutes. Meanwhile, soak gelatin in 6 tablespoons cold water.

When apples are cooked, add dissolved gelatin and stir until completely mixed. Add nuts and lemon juice; mix again. Pour the candy into a shallow pan and let it cool completely. Cut into small squares or rectangles and roll in confectioners' sugar. This amount should make around 4 dozen pieces.

SWISS APPLE TART

One 9 in. pastry case, pre-baked and in its pan
5–6 firm, tart, dessert apples
¼ cup butter
1 teaspoon vanilla essence
3 small eggs
½ cup sugar
1¼ cups cream
Sifted confectioners' sugar, to serve

Peel, core and slice apples thickly. In a large, heavy frying pan, simmer apple slices in foaming butter until they are golden and just tender but not mushy, turning them over gently once or twice with a large spatula. Sprinkle with vanilla and, using the spatula, carefully transfer the slices on to the pre-baked pastry case in a neat, even layer.

Beat eggs with sugar until thick and lemon-colored. Add cream and mix well. Place pastry pan on a cookie sheet. Cover apples with egg mixture, pouring it in quite slowly over the back of a wooden spoon to avoid disturbing apples.

Return tart to the oven for 30 minutes, or until custard has set, no longer trembles when is pushed, and is lightly colored on top. Cool to lukewarm and dust with sifted confectioners' sugar before serving.

Serves 6–8

APPLE JELLIES

6 lbs apples
1 lemon
2 cups sugar per 2½ cups of juice

Chop up apples but do not peel or core them. Finely grate rind of the lemon, sprinkle it over the apples in a preserving pan and cover with about 2½ quarts of water. Bring to the boil and cook until apples are very soft and pulpy. Tip the contents of the pan into a jelly bag and allow to drip dry, or leave the apple pulp to drain in a fine strainer.

Measure juice and weigh out 2 cups of sugar per 2½ cups of juice.

In the rinsed-out pan, combine the apple juice with the sugar and strain in the juice of the lemon. Bring to the boil slowly, stirring frequently, until sugar has dissolved. Then boil rapidly until setting point is reached, skimming scum from the surface.

Cool slightly, pot and cover.

FRESH APPLE RELISH

1 lb apples
3 tablespoons sugar
1 tablespoon crumbled dried mint
Cider or white wine vinegar

Peel the apples and grate them coarsely into a bowl containing the sugar, mint and about 3 tablespoons vinegar.

Toss lightly but thoroughly until well mixed (this will prevent the apples discoloring).

Taste and add more sugar or mint if liked, and if necessary pour in a little more vinegar so that apples are thoroughly moistened.

APPLE CHUTNEY

4 lbs peeled, cored and quartered apples
1 lb seedles raisins, chopped
2 cups dried dates, pitted and chopped
4 cups brown sugar
2 quarts white vinegar
3½ tablespoons mustard seed
3½ tablespoons ground ginger
1 tablespoon salt
½ teaspoon cayenne pepper, or to taste
2 garlic cloves, peeled and very finely chopped

Combine all the ingredients in a preserving pan and mix well. Stir constantly over low heat until sugar has dissolved. Then bring to boiling point and simmer for between 1½ and 2 hours, stirring at an increasing rate as chutney thickens to prevent it scorching on the bottom of the pan.

Pour into a bowl and leave until the following day before potting and covering tightly.

For a darker color, use malt or cider vineger.

Fills about 7 jars

APPLE CIDER SAUCE

2 tablespoons butter
4 tablespoons flour
2½ cups apple cider, warmed
½ teaspoon ground cloves
½ teaspoon ground ginger
Honey or brown sugar

Melt the butter in a saucepan over gentle heat. Mix in the flour and cook, stirring, for 1–2 minutes. Slowly stir in the apple cider and add cloves and ginger. Bring to the boil, still stirring, then remove from heat. Sweeten with honey or brown sugar.

APPLE AND GREEN TOMATO CHUTNEY

1 lb green tomatoes
1 lb apples, quartered and cored
1 lb onions
1 cup golden raisins
1⅓ cups brown sugar
2 tablespoons salt
1 tablespoon ground ginger
1 teaspoon cayenne pepper
½ teaspoon ground cloves
2¼ cups brown malt vinegar

Put tomatoes, apples, onions and raisins through a mincer, together with brown sugar, and put in a preserving pan. Blend salt and spices smoothly with some of the vinegar. Mix with remaining vinegar, pour into preserving pan and mix well.

Place pan over low heat and stir until sugar has melted and mixture comes to boil. Then boil gently for about 1 hour, stirring more and more frequently as chutney thickens to prevent it catching on the bottom of the pan. Cool.

Pot and cover.

Fills about 4 jars

OLD-FASHIONED APPLE BUTTER

6 lbs apples
5 cups sweet cider
Sugar
1½ teaspoons ground cinnamon
½ teaspoon ground cloves

Rinse apples and chop them up. There is no need to peel or core them.

Put them in a preserving pan with cider and barely enough water to cover. Bring to boiling point and simmer the apples until soft and pulpy, stirring occasionally.

Rub contents of pan through a strainer, discarding peels, pips and other debris.

Weigh pulp and return it to rinsed-out pan. Weigh out 1½ cups sugar for every 2 cups of apple purée.

Bring purée to simmering point, stirring frequently. Add prepared sugar and spices. Stir until sugar has dissolved, then simmer, stirring regularly at an ever-increasing rate as purée thickens to ensure that it does not catch on the bottom of the pan. The butter is ready when no surplus liquid can be seen collecting around the edges of a small mound of purée which you have spooned out on to a saucer.

Pot and cover.

BREAKFAST APPLE SHAKE

2 tart apples, peeled and cored
1 cup milk
Pinch ground cinnamon
1 tablespoon honey
Ice cubes (optional)

Chop the apples and put in the bowl of an electric
blender with the milk, cinnamon and honey.

Serves 2–3

Note: Skimmed milk can be substituted for full fat milk to increase
the nutritional value of this drink.

APPLE AND ALMOND PACK

This is good for rough skin (on elbows and knees, for instance), and can also be used as a face pack for dry skin.

Peel and grate a firm apple. Add about the same amount of ground almonds, and mix in milk (or cream) until you have a slightly gritty but spreadable paste. Spread the paste on your elbows, knees (or anywhere else with less than smooth skin) or face, and leave to dry. Rub off from arms and legs – preferably over a bath – or wash off from your face with warm water.

APPLE-BASED REMEDIES

A diet of grated apple, in moderation, will often clear up diarrhea in a couple of days. Use slightly unripe fruit, and peel them first.

Cider vinegar has been hailed as a cure for almost every ill under the sun. Most people are justifiably sceptical about this claim, but it is an excellent blood cleanser and helps to eliminate a vast variety of toxins. Up to two teaspoons in a tumbler of water, two or three times a day, will help digestion and promote energy.

OVEN TEMPERATURES

Very low	250°F
Low	275°F
Very slow	300°F
Slow	325°F
Moderate	350°F
	375°F
Moderately hot	400°F
Fairly	425°F
Hot	450°F

NOTES: